The Wright Brothers as Engineers, An Appraisal

and

Flying with The Wright Brothers, One Man's Experience

by

Quentin R. Wald

Published by the Author
1999

The Wright Brothers as Engineers, an Appraisal
and
Flying with The Wright Brothers,
One Man's Experience

ISBN 0-9706075-0-4

Library of Congress Catalog Card Number: 99-94954

Published by
Quentin R. Wald
102 Cape George Road
Port Townsend Washington 98368

Table of Contents

Part One

Part Two

Illustrations

Preface

Having inherited a collection of photographs of Wright airplanes and the men who flew them, mostly taken by my father, Charles Wald, it was in my mind for many years that I should find a way to make them public and should also put on the historical record the fact that as an employee of the Wright Company he made many flights with a Wright float plane the year before the Wrights are generally credited with first having attempted water based flying. In the course of thinking about these matters I became more and more engrossed in trying to understand the development of Wilbur and Orville Wright's understanding of the problem of flight. In their collected papers, published in 1953 (ref. 7), there is much evidence from which it is possible to trace the progress of their thinking. Looking at their work from the point of view of an aeronautical engineer, it became apparent to me that there has been little adequate discussion of their work as rational engineers in the light of modern aeronautical knowledge. I am now attempting two things, an appraisal of their work as engineers in developing the airplane and an anecdotal and historical account of my father's experience as a pilot for the Wright Company illustrated with some of his photos. I am not sure that the two efforts are entirely appropriate between the same covers, but they have both grown out of my nearly unique circumstances with respect to the material.

I want to thank Ken Hyde who made available to me copies of the correspondence between my father and Orville Wright, material that he found in the collection of Wright papers in the Library of Congress. These letters, whose existence had not been known to me, enabled me to piece together the odd on-again, off-again history of my father's employment by the Wrights and bring the story a bit more to life.

The photographs, Figures 2, 3, 7, and 21, are from the Wright collection in the Dunbar Library of Wright State University, Dayton, Ohio. All other photos are from the collection of the author.

Part 1
The Wright Brothers as Engineers, an Appraisal

Introduction

In the last few years much has been written about the Wright brothers and their invention of the airplane. There have also been some very thorough and competent engineering studies of the Wright "flyer", their first successful powered machine which flew in 1903 (Ref. 9). However, there has not been a clear account of their thinking, their research and their experiments which lead to that successful flight. The following is a brief critical account of that technical endeavor illuminated by modern understanding of aerodynamics and flight dynamics.

Until a few decades ago it was the common impression that the Wright brothers were a couple of middle-western bicycle builders who managed to build a box-kite-like machine that actually flew, thereby initiating the development of aviation. Only in recent decades, following the publication of most of their papers in 1953 (Ref. 7) and subsequent engineering studies of their efforts, has it become apparent to more and more people that they were superb investigators of the problems of designing a flying machine, and that they researched the problems and solved them logically and sequentially as needed in building and flying their machines. In their wind tunnel experiments they showed themselves to be meticulous and ingenious experimenters. In their development of a method of designing propellers they showed considerable insight into some theoretical problems.

The difficulty of putting oneself in the mental landscape of a past time makes it difficult to evaluate fairly the insights and weigh the shortcomings of the pioneer thinker. It is too easy to see his shortcomings and blind spots and, on the other hand, the enthusiasm of the evaluator may slip too easily into unbridled admiration. In discussing the work of Wilbur and Orville Wright I am trying to achieve a balanced point of view. Some things would have come to them more quickly and would have been better understood if they had had a scientific education. Yet they accomplished in a remarkable four years of less than full time work the successful manned flight that eluded their near contemporaries Hiram Maxim and S. P. Langley, both well educated and with far greater access to financial resources.

Let me state what are, in my opinion, the great accomplishments of the development program undertaken by the Wright brothers. (1) Identification of control as the primary unsolved problem. (2) Realization that an airplane must bank in order to turn in a horizontal plane and the invention of a means of lateral control which came to be known as "wing warping". (3) Recognition of the problem of "adverse yaw" and invention of a practical solution in the form of a rudder with action coupled to the wing warping control. (4) The carrying out of a program of wind tunnel experiments which resulted in the measurement of the aerodynamic characteristics of something like two hundred model lifting surfaces. The simplicity and effectiveness of the experimental equipment was remarkable. These experiments enabled them to choose efficient geometric characteristics, including area, aspect ratio, airfoil section and tip shape for the wings of their successful 1902 glider and the powered machine of 1903. (5) The truly remarkable design of efficient propellers entirely from theoretical considerations. In this they seem to have been the first to combine, in a primitive way, blade element theory with momentum theory. (6) Probably unique in the history of invention is their successful mastering of the art of using their invention at the risk of their lives as they developed it.

The Prehistory of Flight

Consider the state of the problem of flight and the technology of the time. From ancient legends until recent times men have thought to fly by attaching wings to their arms and flapping in imitation of the birds. Such efforts were based on little understanding of the forces involved and the energy required. During the nineteenth century a great many inventors and engineers experimented with models or full size flying machines. The work of many of these is described in Ref. 2 but here I will only mention a few of the most important.

The first serious advance toward the concept of the airplane was made by Sir George Cayley around 1810 when he proposed a fixed wing glider with a vertical fin and a horizontal stabilizing surface at the rear. He built many model gliders of this general configuration. In 1842 William Samuel Henson applied for a patent for an "aerial steam carriage" which embodied most of the essential features: a cambered wing surface, fuselage, movable tail surfaces for pitch and yaw control and twin pusher propellers. The improbable man-powered ornithopter was left far behind and we see a proposal for a well engineered wing structure and a primitive form of the screw propeller driven by an engine.

In 1871 in Paris Alphonse Penaud demonstrated a flying model which had a fixed wing of cambered section, stabilizing surfaces at the rear, a pusher propeller and rubber band power similar to the arrangement of simple modern flying models. The importance of his models was the demonstration of aerodynamic stability of an uncontrolled model having suitable tail surfaces and wings with a dihedral angle.

During the 1890's Otto Lilienthal in Germany made many glides with what we now call a hang glider. In these machines there are no aerodynamic controls, the operator applying control simply by shifting his weight. In the United States Octave Chanute, a successful civil engineer, built simple gliders of the hang glider type. His notable contribution was the wire braced biplane forming a light and strong truss. This structural feature was adopted by the Wrights and was the dominant structural configuration for wings in most aircraft from the Wrights into the 1930's. It only became obsolescent with the realization that it was possible to use thicker wings with internal bracing and thereby eliminate the cost in speed or power required to push external bracing of wings through the air.

In a very different line of development from the gliders were the large steam powered flying machines of Maxim and Langley. In England, Hiram Maxim built an enormous steam powered flying machine with a wing span of 107 feet which was confined to a track. In 1894 it broke loose from its track and was wrecked, but if it had ever flown would doubtless have been uncontrollable.

During the 1890's Samuel P. Langley, Secretary of the Smithsonian Institution, built large steam powered flying models up to fourteen feet wing span. Finally, in 1903 he built a full size man-carrying machine which collapsed in a tangle of wreckage when it left its launching track.

It is remarkable that neither Langley nor Maxim contributed a single useful principle to later experimenters and had no discernible influence on the development of aircraft while the humble gliders of Chanute and the experimental work of Lilienthal had a traceable influence on the Wrights and others.

The State of Technology at the Turn of the Century

Before the development of the steam engine, powered flight could only be imagined using human muscle power, which was grossly inadequate. With the advent of successful steam engines and screw propelled ships in the nineteenth century, the conception of man-carrying flying machines became more realistic, but the steam engine remained excessively heavy and inefficient for flight. Powered flight only came within the reach of man with the advent of the internal combustion engine.

When Wilbur Wright began to think seriously about the problem of flight, the railroad and the steamship were common modes of transportation, but electricity was not yet a household convenience and the automobile had scarcely begun its development. Oil lamps were the common lighting and the horse was the everyday mode of transportation. In Germany Nikolaus Otto produced his first successful internal combustion engine in 1876, an engine employing the four stroke cycle still known as the Otto cycle. Otto engines were first produced in the United States in 1878. In 1899 when Wilbur Wright first seriously addressed the problem of flight, there were few compact and reliable internal combustion engines available.

Aerodynamics as a science hardly existed at the turn of the century, although hydrodynamics existed as a quite well developed body of theory. Aerodynamics developed as an extension of hydrodynamics into the problems of lifting surfaces, but only in the years following the pioneering work of the Wrights. In 1907 F. W. Lanchester published in London his "Aerodynamics" which presented seminal ideas in the theory of lifting surfaces, but his work lacked a mathematical exposition and was very difficult to understand. In 1911 Ludwig Prandtl published in Germany a paper which began the quantitative analysis of lifting surfaces. Over the following decade Prandtl and his associates and others developed the theory of lifting surfaces in considerable detail. The theory of flight dynamics, i.e. the theory of airplane motion as distinct from aerodynamics, was first developed by G. H. Bryan in 1904. It is evident that there was little useful theoretical knowledge available to the Wrights at the time of their pioneering work (1899-1903). Of necessity they were dependent upon their own intellectual resources. It is clear from the dates of the basic theoretical developments that they were inspired by the reality of flight and had not been available as useful tools for the development of the first flying machines.

Beginnings

From childhood Wilbur Wright had an interest in the problem of flight. In 1896 when he heard of the death of Otto Lilienthal, the German experimenter, he decided to make a serious effort to contribute something substantial to the problem of flight. "...as my brother soon became equally interested with myself, we soon passed from the reading to the thinking, and finally to the working stage." In 1899 when he was 32 years old, he wrote to the Smithsonian Institution requesting copies of papers pertaining to flight and suggestions for other pertinent technical literature. He had soon read everything of consequence available in English, and had drawn his younger brother Orville into his plans. They had both completed high school and had a successful bicycle manufacturing and repair business in Dayton, Ohio which afforded them some time and modest means to pursue their interest.

Of the two brothers, Wilbur was the original thinker whose vision started them on the path of invention. Orville was soon committed to the project and ever after they were careful to present a united front to the world, always using the word "we" in describing their activities. Orville was an ingenious mechanical designer and was recognized by their pupils as a superbly skilled pilot.

Wilbur Wright's Perception of the Problem

Wilbur Wright's conception of the problem of flight is outlined in a talk entitled "Some Aeronautical Experiments" which he presented before the Western Society of Engineers on September 18, 1901 in Chicago, two years before the historic first powered flight. I quote some excerpts:

> The difficulties which obstruct the pathway to success in flying machine construction are of three general classes: (1)Those which relate to the construction of the sustaining wings. (2)Those which relate to the generation and application of the power required to drive the machine through the air. (3)Those relating to the balancing and steering of the machine after it is actually in flight. The person who merely watches the flight of a bird gathers the impression that the bird has nothing to think of but the flapping of its wings. As a matter

> of fact this is a very small part of its mental labor. Herr Otto Lilienthal seems to have been the first man who really comprehended that balancing was the first instead of the last of the great problems in connection with human flight. He began where others left off, and this saved the many thousands of dollars that it had theretofore been customary to spend in building and fitting expensive engines to machines which were uncontrollable when tried. He built a pair of wings of a size suitable to sustain his own weight, and made use of gravity as his motor.
>
> We figured that Otto Lilienthal in five years of time had spent only about five hours in actual gliding through the air. The wonder is not that he had done so little, but that he had accomplished so much. It would not be considered safe for a bicycle rider to attempt to ride through a crowded city street after only five hours' practice, spread out in bits of ten seconds each over a period of five years...

It is clear that very early in his thoughts of flying Wilbur Wright had laid aside the idea of beginning with the construction of a large powered man-carrying machine such as Henson had proposed and Langley and Maxim had actually built. He also set aside experimentation with small models which could not illuminate the problems of control. He would begin as a follower in the direction established by Chanute and Lilienthal, experimentation with simple unpowered man-carrying gliders. With these simple machines he would try to solve what he considered to be the essential problem, control, and its attendant difficulty, learning to use that control in actual flight. However, Wilbur Wright made a great departure from the practice of these forerunners. He abandoned the very limited method of control by shifting the operator's weight and sought to provide purely aerodynamic control. Sustained powered flight would come later when control was understood and flying skill had been acquired. In his view of the problem of flight he proved himself most perceptive, but could hardly have imagined that his concentration on the control problem would result in powered flight only four years later.

Earlier experimenters and thinkers on the problem of flight had realized that the solution lay in the provision of an adequate expanse of lifting surface propelled through the air by mechanically driven screw propellers, the latter no doubt suggested by the success of screw propulsion for steam powered ships. The provision of a rudder for steering was also a natural transfer of ideas, but the necessity for lateral (roll) control and control in pitch (nose up or nose down) seems to have been much less evident. Some experimenters provided a horizontal surface behind the wings which would serve as a "rudder" for pitch control. Experimenters with models all came to realize that setting the wings at a dihedral angle provided lateral stability, but none before Wilbur Wright seem to have thought it necessary to provide roll control. The Wrights seem to have been distrustful of the idea of a stable craft, thinking that its response to gusts of wind would take control of the machine away from the operator. Perhaps as a result of their experience with bicycles, they felt that stability was unnecessary and should be foregone in deference to effective control.

The Forward Elevator

It is at first quite difficult to understand why the Wrights chose to place a control surface ahead of the wing when several experimenters, including Cayley and Penaud, had already demonstrated longitudinal stability with a horizontal surface behind the wing. The problem is further muddied by various biographers, good historians but with little understanding of mechanics, who have written elaborate discussions of the Wrights' analysis of the stability and control problem, intertwining their own confusion with that of the Wrights. From the writing of the brothers themselves, however, especially Wilbur Wright's talk before the Western Society of Engineers in 1901, there is ample indication of their initial thinking about stability and control.

Although they were well aware that longitudinal stability could be attained, as it had been in many model gliders, the Wrights certainly did not understand the mechanics of longitudinal stability. Stability is the property of tending to return to an equilibrium condition after a disturbance. It can best be understood by consideration of the changing moments that arise when the machine is rotated nose up or nose down. The Wrights never, so far as I know, used the word "moment" in connection with longitudinal stability. They always thought in terms of the center of pressure and its travel from an equilibrium position at the center of gravity of the machine. Their objective was always to keep the center of pressure close to the center of gravity regardless of the angle of attack of the wings. If achieved, this would be a condition of neutral stability; that is, after a disturbance of the attitude of the machine, there would be no tendency to restore the initial equilibrium nor to depart further from it.

The following are some excerpts from the lecture before the Western Society of Engineers.

> After much study we finally concluded that tails were a source of trouble rather than of assistance; and therefore we decided to dispense with them altogether....Then the method of control used by Lilienthal, which consisted in shifting the body, did not seem quite as quick or effective as the case required; so, after long study, we contrived a system consisting of two large surfaces on the Chanute double-deck plan, and a smaller surface placed a short distance in front of the main surfaces in such a position that the action of the wind upon it would counterbalance the effect of the travel of the center of pressure on the main surfaces. Thus changes in the direction and velocity of the wind would have little disturbing effect, and the operator would be required to attend only to the steering of the machine, which was to be affected by curving the forward surface up or down. ...In deeply curved surfaces the center of pressure at 90 degrees is near the center of the surface, but moves forward as the angle becomes less, till a certain point is reached, varying with the depth of curvature. After this point is passed, the center of pressure, instead of continuing to move forward, with the decreasing angle, turns and moves rapidly toward the rear. The phenomena are due to the fact that at small angles the wind strikes the forward part of the surface on the upper side instead of the lower, and this part altogether ceases to lift, instead of being the most effective part of all, as in the case of the plane.

> Our peculiar plan of control by forward surfaces, instead of tails, was based on the assumption that the center of pressure would continue to move farther and farther forward, as the angle of incidence became less, and it will be readily perceived that it would make quite a difference if the front surface instead of counteracting this assumed forward travel, should in reality be expediting an actual backward movement.

The possibility admitted by Wilbur Wright in the last paragraph is, in fact, the reality. It is true that, starting at an angle of 90° the center of pressure (CP) is at the mid chord of the wing and that, as the angle is decreased, separated flow is replaced by smooth flow and the CP moves forward, but all this is at angles not suitable for normal flight. At angles less than the stalling angle, where the machine normally flies, the CP of a cambered airfoil moves steadily *backward* as the angle of attack is further decreased, resulting in an increasing nose down moment. If it were true that the CP moved forward with decreasing angle of attack, the wing would in fact be stable, that is, as the angle decreases, there would be an increasing tendency to resist the nosing down. By their own reasoning, their forward elevator was tending to cancel a movement of the CP which would have been a stable condition. In actual fact, the elevator was aggravating an unstable condition.

The assumption that the center of pressure of a lifting surface moves continuously forward as the angle of attack is decreased from 90° did not originate with the Wrights, but was apparently common in the aeronautical literature of the time. It is apparently based on the intuitive (and approximately correct) assumption that the CP is at the mid chord when the surface is normal to the air stream and the belief that it must be at the leading edge when the angle of attack is zero because only the leading edge of a thin flat board would seem to be exposed to the pressure of the oncoming stream. The latter assumption is false and the CP of a thin uncambered foil is practically stationery at approximately the 25% chord point at all angles of attack below the stalling angle. For a cambered surface, the CP at the angle of maximum lift is somewhere behind the 25% point and moves steadily backward as the angle decreases. (See Fig. 1).

The Wrights inadvertently produced aircraft which were very unstable in pitch. Eventually they seem to have come to realize this, but by that time they had learned to fly the unstable machines and felt that the great effectiveness of their forward elevator in controlling a possible dive was a feature which could and did save them from disaster in extreme situations. The serious instability and the difficulty of flying these machines is evidenced by early movies showing persistently undulating flight. The configuration of their early machines was designed under the influence of an erroneouÛ assumption regarding the travel of the center of pressure of wings. If better information on airfoil characteristics had been available to them, the Wrights might never have adopted the forward elevator configuration which they eventually abandoned. Not until 1905 did they conduct a few limited experiments in which they actually determined the position of the center of pressure.

It may be asked why we only discuss the action of the forward elevator as the angle of attack of the wing *decreases*. The answer is that that is what the Wrights were worried about. An uncontrolled *increase* of angle of attack would result in a stall which is followed by the dropping of the

CENTER OF PRESSURE IS CONVERTED FROM MOMENT COEFFICIENT DATA OF REFERENCE 1.

$$x/c = (x/c)_{ac} - C_M/C_L$$

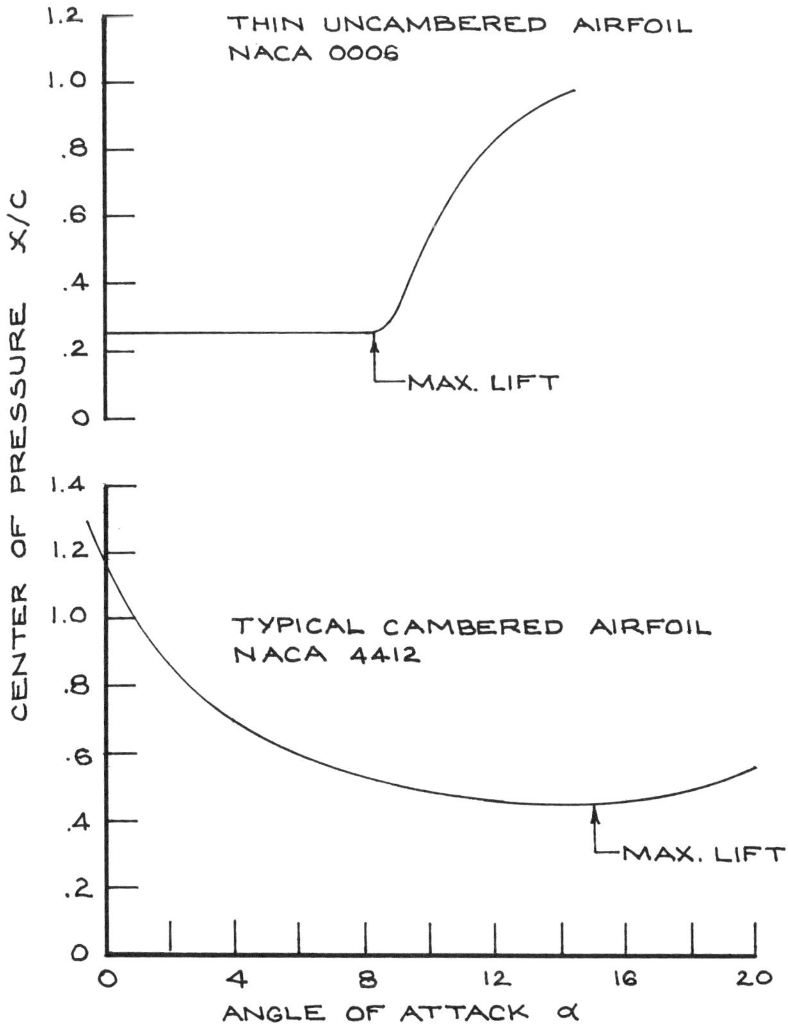

CENTER OF PRESSURE OF TYPICAL AIRFOILS

FIG. 1

nose. This is what had killed Lilienthal and others; consequently the Wrights were not worried about nose up instability, but the subsequent dive. The forward elevator in fact made the Wright machines unstable in pitch, nose up or nose down, but they felt secure in the knowledge that their forward placement of the control surface, quite apart from the matter of stability, provided a sure and effective *control* in all flight conditions. Seen against the horizon, it also provided a reassuring visual indication of the attitude of the machine.

Much has been made of the forward elevator configuration used by the Wrights as a clever safety feature which obviated the possibility of violent nosing down should a stall occur in the course of their learning to fly their gliders. This configuration is said to result in a relatively mild sinking without the abrupt pitching down usually associated with a stall. The fact is that this behavior of the Wright gliders of 1900-1902 is more the consequence of the position of the center of gravity much too far back for the machine to be stable rather than to the geometry of the control surfaces. An airplane with the now conventional tail configuration would also display this "mushing" behavior if the center of gravity were moved to a point well behind the neutral point. (See the section on the evolution of the longitudinal stability of Wright airplanes, page 45, for an explanation of neutral point).

Lateral and Directional Control

It is interesting that the gliders that the Wrights built in 1900 and 1901 had no vertical surfaces whatsoever, lacking what we call a rudder. How did they expect to maneuver in the horizontal plane? Again I quote Wilbur's 1901 lecture.

> The lateral equilibrium and the steering to right or left was to be attained by a peculiar torsion of the main surfaces, which was equivalent to presenting one end of the wings at a greater angle than the other.

The "peculiar torsion of the main surfaces" is the ingenious system of lateral control invented by the Wrights and described in their basic patent. It was used in all their machines and consisted of a means of twisting the wings by pulling on wires which formed a part of the system of wing trussing, a control device eventually to be known as wing warping. The front wing spars and struts were wire braced to form a rigid truss. The rear spars were similarly braced in the center section of the wings, but the outer portions of the rear spars were arranged to bend up or down by the action of diagonal control wires, thereby twisting the outer portions of the wings. The twist increased the angle of attack on one wing and decreased it on the other to produce a like difference of lift. It is effective roll control and mechanically practical for thin trussed biplane wings. It was employed in many machines in this country and in Europe into the beginning of the first World War. Glenn Curtiss and others evaded the Wright patent by using separate small controllable wings or "ailerons" between the main lifting surfaces or ailerons in the form of hinged panels at the trailing edges of the wings, as all airplanes do today.

There seems to be some mystery how the system of lateral control, "wing warping", was also to provide "steering to right or left". One might be tempted to think that, trying to solve one problem at a time in their typical methodical way, they did not expect to make turns with that first glider, but only expected to glide near the surface of a hill in the direction in which the machine was initially pointed. However, Wilbur does refer to "steering to right or left".

He had realized that in order to turn, a lateral force toward the center of the intended turn (a centripetal force) must be produced. This is done by birds and all airplanes by banking, or rolling. The wings are inclined so that a component of the lift becomes a lateral force. This will initiate a sideward acceleration which there must be in flight along a curved path, but it is also necessary to rotate the machine about a vertical axis in order to point it in the direction of the intended turn. This action is provided by a fixed vertical fin or a rudder at the aft extremity, a device not incorporated in the original Wright gliders.

In many modern aircraft it is possible to initiate a turn by applying roll control, that is by banking, without applying the rudder, but only in craft with good directional stability which is provided by a large vertical tail fin. In such a craft, the lateral motion induced by rolling in the direction of the intended turn results in a force on the fin which rotates the machine about its vertical axis. With no vertical surface at all, it is not clear how Wilbur Wright expected to produce a turn with the warping control alone. It appears that he did not realize that a proper turn requires a combination of roll and yaw.

Wing warping was found to be a satisfactory control for raising one wing tip after the disturbance of lateral equilibrium by a gust. However it was found that from time to time, probably when the control was applied longer than momentarily, the glider would turn toward the wing which the pilot was trying to raise, and the wing would then fail to rise. The Wrights soon realized that the wing which presented an increased angle of attack was subject to increased drag as well as increased lift. The increased drag of one wing when roll control is applied, an effect now known as adverse yaw, tended to rotate the machine about a vertical axis. The rotation reduced the speed of that wing through the air and thereby caused a loss of lift rather than the desired increase. The consequence was a violent uncontrollable rotation about the vertical axis and dropping of the wing which the pilot was trying to raise, a disastrous motion they called "well digging". At the end of the flying season at Kittyhawk in 1901 the Wrights were quite discouraged by this unruly and dangerous behavior of their glider.

The problem was solved with the glider they brought to Kitty Hawk in 1902. The first solution to the problem of the uncontrollable rotation was to add, behind the wing, a fixed vertical surface, a fin which would resist the tendency of the machine to yaw, or rotate about a vertical axis. The control of the machine was much improved and its gliding performance was vastly better than that of the earlier gliders. They were encouraged to make bolder maneuvers, but then found that the problem reappeared. They decided to make their fixed fin a movable rudder which could be activated to stop an undesired yawing motion, but were reluctant to burden the pilot with an additional control. This difficulty was avoided by connecting the rudder control wires to the wing warping wires. Thereby the application of roll control automatically brought into play the necessary rudder action to compensate for the adverse yaw. At this point they had completely solved the fundamental problem of control of an airplane, having provided effective aerodynamic control for rotation of the machine about each of the three axes. No one had previously accomplished this. Their basic patent covered the control system they had developed and proven in flight. It covered nothing more, nor was more necessary.

Having described the solution of the control problem in 1902, we turn our attention back to other serious problems that troubled them in 1900.

Lift Less Than Expected

In 1900 the brothers prepared for their first experiments with a glider. They first planned to build a machine which would provide the necessary lift to support a man and the structure at 18 miles per hour. Using tables of forces on cambered surfaces compiled by Otto Lilienthal, they calculated the required wing area to be about 200 square feet. They carefully selected a site where there were flat expanses and wind velocities high enough that the weight of a man-carrying machine could be supported with wings of modest size tethered by a tow rope attached to an elevated fixed point, or as a kite. At Kitty Hawk, North Carolina where they planned to conduct their gliding experiments, winds of 16 to 25 miles per hour were common. Adequately long lumber for the wing spars as originally planned turned out not to be available and the glider as built had a wing area of 165 square feet which meant that the desired lift could be expected at about 21 mph.

2. Wilbur Wright Flying the 1902 glider

As it turned out, the expectation of long practice flying at the end of a tow rope extending horizontally from a tower proved impractical. Flying the machine as a kite with no man aboard, the lateral control by wing warping (by means of control lines in the hands of the operator on the ground) was found to be entirely satisfactory. Flown as a manned glider, the longitudinal control with the forward horizontal "rudder" proved very satisfactory. But both as a kite and as a glider, the lift was much less than expected and they were led to doubt the reliability of Lilienthal's tables.

The disappointing lifting performance led the Wrights to doubt the reliability of the information on lifting surfaces available to them, their principal source being Lilienthal's tables. Lilienthal tested lifting surfaces using a rotating arm to move the models through the air and also made many measurements using natural winds. The rotating arm and natural wind experimental techniques are subject to serious difficulties and possible errors and were abandoned early in the twentieth century in favor of the use of wind tunnels. However, Lilienthal's experimental technique and presentation of the results were exemplary. He presented his results as ratios of the lift and drag forces on various surfaces to the force on the surface when disposed normal to the wind, which is to say as a dimensionless coefficient. The presentation cannot be faulted, but in order to obtain useful quantitative results, a user of his tables and graphs would have to know the force on the surface when it is oriented normal to the wind. In this calculation Lilienthal seems to have committed a serious oversight which appears to have misled the Wrights. In order to appreciate this difficulty it is necessary to examine briefly the way in which aerodynamic forces were expressed mathematically by Lilienthal and others of his time.

The Lift Equation Used by the Wrights and its Application

Through the nineteenth century, the force on a flat plate set normal to the wind direction was conventionally expressed by the formula

$$D = k\,S\,V^2 \qquad (1)$$

where k was referred to as Smeaton's coefficient after John Smeaton, an eighteenth century engineer who had measured wind forces. The coefficient k was usually assigned the value .005 when S was in square feet, V in miles per hour and D in pounds.

There are some obvious flaws in the use of this relation. It takes no account of variations of air density and the coefficient would actually have somewhat different values for surfaces of various shapes. Modern measurements indicate that for standard sea level density the coefficient in the Smeaton equation is properly .0030 for discs and square plates and rectangular surfaces of moderately low aspect ratio. (See note below.) The formula was adequate for the purposes of early experimenters who worked near sea level, but the customary value of the coefficient k was seriously misleading for experimenters requiring reasonably accurate values for the purpose of designing lifting surfaces.

Early experimenters in aerodynamics including Otto Lilienthal and the Wrights found it convenient to express the lift or the drag of a lifting surface as the ratio of the force to the force on a flat plate of the same area set at right angles to the wind. That is, the force given by the Smeaton formula was multiplied by a suitable lift or drag coefficient which was determined by experiment. Consequently lift was expressed as:

Note:

The force on a flat plate set normal to the wind is given in Sighard F.Hoerner's book "Fluid-Dynamic Drag", (Ref. 6). He states that the drag coefficient C_D is 1.17 for discs and square plates and increases only slightly for rectangular plates of aspect ratio up to about ten. Equating the force by the Smeaton formula to the force given by the equation now in use,

$$D = k\ S\ V_{mph}^2 = C_D\ \tfrac{1}{2}\ \rho\ S\ V_{fps}^2$$

here $\rho = .002378$, the standard sea level air density in slugs / ft^3 or lb sec^2 / ft^4.

Solving the above equation for the Smeaton coefficient k,

$$k = C_D\ \tfrac{1}{2}\ \rho\ (V_{fps} / V_{mph})^2$$

$$= 1.17 \times \tfrac{1}{2} \times .002378 \times (5280/3600)^2 = .0030$$

$$L = C_L\ k\ S\ V^2 \qquad (2)$$

where C_L is the lift coefficient. (In modern aerodynamics a different expression is used and it includes the air density as a factor. To avoid confusion it is necessary to realize that C_L now has a different meaning than it has in the above relation).

Lilienthal seems to have misled his readers by frequently quoting the relation (2) including the traditional value of the Smeaton coefficient. (In metric units the English value .005 became 0.13.) It is surprising that, meticulous investigator that he was, he seems never to have attempted to verify the received value of the coefficient. He even measured wind velocity by observation of the deflection of a spring which opposed the wind force on a flat plate. To relate the spring force to wind velocity he accepted the Smeaton value of the coefficient. His wind velocity measurement would have been seriously in error, but this did not affect the validity of his tables and graphs which presented lift and drag forces as ratios to the force when the surface is oriented normal to the wind. Unfortunately, users of his tables were misled by his repetition of the traditional value of the Smeaton coefficient which they would use in applying the tables.

In experiments with gliders in 1900 and 1901 the Wrights, having found the lift and drag to be significantly less than they had anticipated, apparently suspected that an incorrect value of Smeaton's coefficient was the cause of the discrepancy. This was undoubtedly the cause of a large part of the problem, but there were other factors. In transmitting the tables to the Wrights, Octave Chanute had failed to indicate the plan form of the models tested by Lilienthal. The importance of aspect ratio was not appreciated and the very different plan form and low aspect ratio employed by the Wrights in their first gliders resulted in serious differences in performance.

Wind Tunnel Experiments

At first discouraged by the deficient lift of their gliders, the Wrights determined to rely only on their own resources. In December, 1901 they constructed a very simple wind tunnel with which to measure the forces on lifting surfaces. A wind tunnel is a tube through which a current of air is made to flow. Model lifting surfaces or other bodies are mounted in the air stream and the aerodynamic forces on them are measured by a suitable weighing device referred to as a balance. The Wrights were by no means the first to conduct experiments in a wind tunnel, but they were the first to make extensive measurements on lifting surfaces in a wind tunnel.

Their wind tunnel was a wooden tube sixteen inches square and about six feet long. Air was forced through it by a fan driven by the one cylinder gas engine in use in their shop. The swirling air flowing from the fan was straightened, to a degree, by flat vanes immediately downstream of the fan. The resulting flow must have been extremely turbulent and rather unsteady. It is remarkable that, through great care, they were apparently able to get consistent measurements of forces on model lifting surfaces. It is curious that they did not avoid the turbulence generated by the fan by placing it at the exit from the tunnel to draw the air through rather than placing it at the inlet.

The Wrights, in their wind tunnel experiments, did not measure forces directly but, by an ingenious mechanical arrangement, measured the ratio of the lift on a surface to the drag of plates set normal to the wind. Using this device, they were measuring not the force, but C_L directly. There were some great advantages in this scheme, but then it became necessary to know an accurate value of the Smeaton coefficient k in order to apply the experimental results to find the lift of a wing by means of equation (2). The results were then only as good as the assumed value of k. Surprisingly, the Wrights, like Lilienthal, made no attempt to obtain by independent experiment a value of the coefficient.

It would have been the obvious and natural thing to measure lift directly and not as a ratio of forces. Perhaps suggested by the lift formula (2) in which C_L is the ratio of lift to the force on a flat plate normal to the stream, they constructed, largely from old hacksaw blades and bicycle spokes, a simple mechanical device (Fig. 3) which measured this ratio. In choosing this method of measurement they happily avoided some difficulties which were perhaps greater than they realized. It would have been very difficult to measure accurately the velocity of a steady air stream at the low velocity, something like 25 or 30 m.p.h., that they were able to produce. The difficulties would have been compounded by the unsteady flow that must exist in a stream driven by a one cylinder engine. With their ratio balance, they had no need to measure wind velocity and were untroubled by fluctuations of velocity which affected the model being tested and the reference plates in equal measure, leaving steady the measured ratio of forces.

The mechanical principle of the balance is shown schematically in Fig. 4. It consists of two linkages, ABB′A′ and ACC′A′, each having three members. A and A′ are fixed pivot points common to both linkages. The two linkages pivoting about A and A′ describe two parallelograms. The lifting surface being tested is mounted on BB′ so that its lift L is to the right as indicated. A set of flat plates of total area

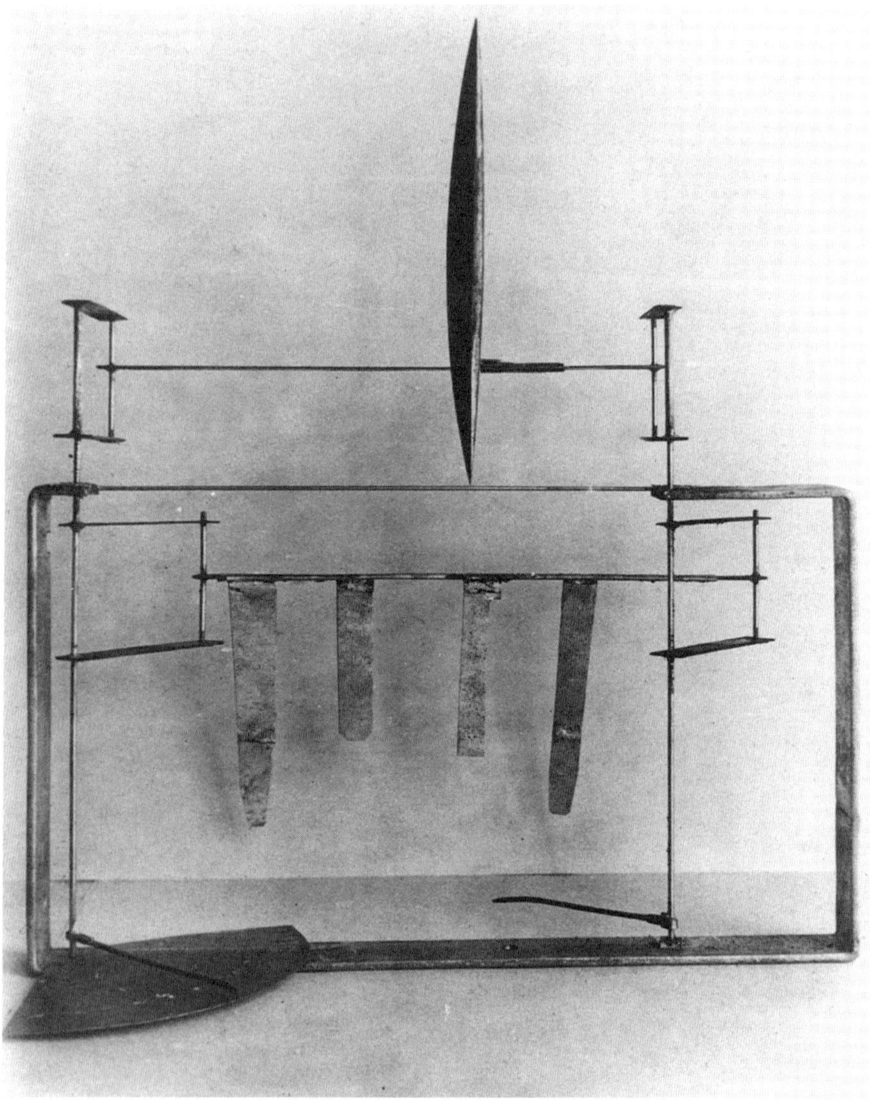

3. The Wind Tunnel Balance

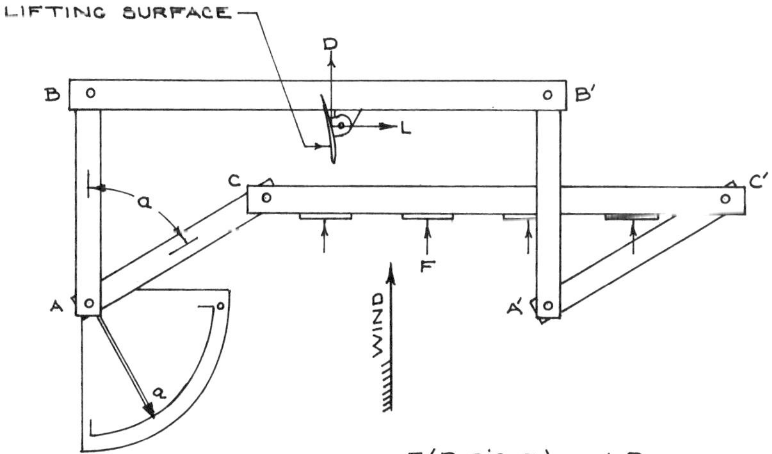

$$F(R \sin \alpha) = LR$$

$$\therefore L/F = C_L = \sin \alpha$$

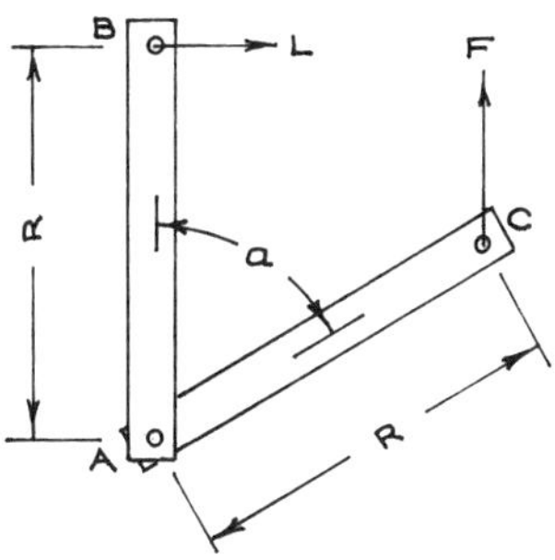

PRINCIPLE OF THE WRIGHT LIFT BALANCE - 1901

FIG. 4

equivalent to that of the lifting surface is mounted on CC′. A friction device is provided at A which holds AB and AC at a constant angle a to each other but leaving them free to rotate together about A. The angle a is indicated on the protractor. When some arbitrary angle is set and the wind is turned on, the lift force tends to rotate the linkages to the right while the force F on the flat plates tends to rotate them to the left. Each time a new model wing is mounted and for each angle of attack, the wind must be turned on and the angle a adjusted so that AB and A′B′ float parallel to the wind. This assures that the drag force D, which is parallel to the wind, exerts no moment about the pivots A and A′.

When the device has been adjusted, the forces are as indicated in the simplified lower diagram. Since the linkages are unconstrained in rotation about A and A′, the clockwise moment due the lift force L must be balanced by the opposing moment of the force F on the plates.

$$F\,R \sin a = L\,R$$

hence the desired ratio is just

$$L/F = C_L = \sin a \tag{3}$$

This is not an automatically reading balance. For each test condition it must be manually balanced by adjusting the angle a with the wind on. It is remarkable that the Wrights investigated about 200 different lifting configurations in the month of December, 1901. Another simpler linkage in the form of a parallelogram was used to measure directly the ratio of drag to lift. With these two measured quantities they had the essential information necessary to evaluate the performance of a lifting surface at each angle of attack.

Models were made of sheet steel cut and hammered to the desired curvature (camber). In many cases the leading edge was thickened and rounded by applying solder or wax to the steel sheet. In this way they were able to test many different airfoil shapes, varying the thickness, leading edge form, camber, and position of maximum camber. Wings of varying aspect ratio and different tip shape were tested. Monoplanes, biplanes and a few triplanes were tested.

After they had conducted wind tunnel experiments, Wilbur Wright was able to deduce a value for the Smeaton coefficient by comparing glider experiments with the wind tunnel data, a remarkably round about way of finding the forces on the flat plates in his wind tunnel which were used to balance the lift forces. The wind tunnel tests yielded directly values of C_L. The glider flights for the same lifting surface shapes, when lift, area and velocity were substituted in equation (2), gave him values of the product C_L x k. Comparing these results, he was able to arrive at a value of the coefficient k. Wilbur Wright's deduction of the value of Smeaton's coefficient is a chain of reasoning subject to many possible inaccuracies including differences between Lilienthal's models and the Wright glider and, not least, the difficulty of measuring the performance of a glider with any degree of accuracy, but he arrived at a reasonably good value. He concluded that k should be about .0032 which happens to be a fair value for a circular plate at standard sea level conditions.

At the conclusion of this work the Wrights had the knowledge which enabled them to select wing configurations that were efficient, something no other experimenter had attained. Not only that, but they were able to compute the necessary wing area to lift the intended load at the desired speed, were able to estimate the gliding performance of a projected design and could calculate the power required to propel a machine. The results were striking. They returned to Kittyhawk in September, 1902 with a glider incorporating many improvements suggested by their wind tunnel tests of lifting surfaces and with a device which they had devised to cure the loss of control which had accompanied vigorous use of the lateral control. The 1900 and 1901 gliders look primitive to us now, their surfaces of crude form and the aspect ratios unsuitable. The 1902 glider, Figure 2, on the other hand, has an impressive rightness about it. The aspect ratio is typical of many modern aircraft, the wing tips have a pleasing and suitable shape and the wing has both the upper and lower surfaces covered to form an efficient airfoil. With the solution of the control problems in 1902, the Wrights possessed the knowledge to build a powered machine whose control characteristics and performance were fairly well known before it flew.

The Wright Propellers

The Wrights concentrated on the problem of control of an unpowered aircraft, confident that when the time came to provide power a suitable engine could be found and propellers could be designed for the aircraft by adapting the technology of marine screw propellers. In both of these anticipations they proved to be mistaken.

They were disappointed and no doubt surprised to discover that, after a century of use, the design of marine propellers was almost entirely lacking a theoretical basis. Propellers for ships had been developed entirely by empirical methods that provided no basis for adaptation of the technology to aircraft. This is not to say that no theory existed, but it was rudimentary and had been of little or no value to the designers of marine propulsion systems.

The Wrights realized that a propeller blade was a lifting surface similar to a wing moving on a helical path. They had very ample data gathered in their wind tunnel experiments to enable them to predict the forces generated by lifting surfaces or, as they called them "aeroplanes", but the problem proved confusing. In an article published in 1908 they wrote:

> ...What at first seemed a simple problem became more complex the longer we studied it. With the machine moving forward, the air flying backward, the propellers turning sidewise, and nothing standing still, it seemed impossible to find a starting point from which to trace the various simultaneous reactions. Contemplation of it was confusing... It was not until several months had passed, and every phase of the problem had been thrashed over and over, that the various reactions began to untangle themselves.

In December, 1902 they made some tests of a 28 inch diameter fan and in February, 1903 they made stationary tests of an 8.5 foot propeller. These experiments no doubt improved their confidence in the theory they had developed, but they were well aware that they had no means of proving the performance of a propeller at flight speeds where conditions would be very different from a stationary test. They had no choice but to rely on their theory and proceed with design and construction of propellers for their first powered machine.

There were two very different rudimentary theories of the action of the screw propeller available to the Wrights. S. Drzewiecki had originated a blade element theory of propellers which considered the blades to be made up of a large number of segments or elements, each acting as an independent airfoil. The breaking up of the blade into many elements is necessitated by the fact that at each radius the speed of the blade and the direction of the forces on it are different. Drzewiecki considered the forward velocity and the rotational velocity of the elements of the propeller blades, but took no account of the motion of the air due to the action of the propeller itself. Fig. 5 depicts a blade element and its component velocities.

The second line of thinking, pioneered by W. J. M. Rankine and William Froude, relates the thrust on the propeller to the movement of air it induces. Essentially, the thrust of the propeller continuously imparts momentum to the column of air through which it moves in accordance with Newton's second law of motion, but momentum theory says nothing about the form of the propeller. A sound physical theory of propeller action could only be built on the combination of these two general principles. The Wrights appear to have been the first to attempt this synthesis, albeit in a primitive way. The resulting theory proved adequate for the design of quite efficient propellers.

The Wrights' understanding of the implications of the momentum theory of propellers is attested by their choice of two large slow-turning propellers which would work on as large a volume of air as possible. This was necessary in order to attain the high efficiency that was absolutely required for flight with a very low powered engine. By crossing the chain drive of one of the two propellers, they were also arranged to

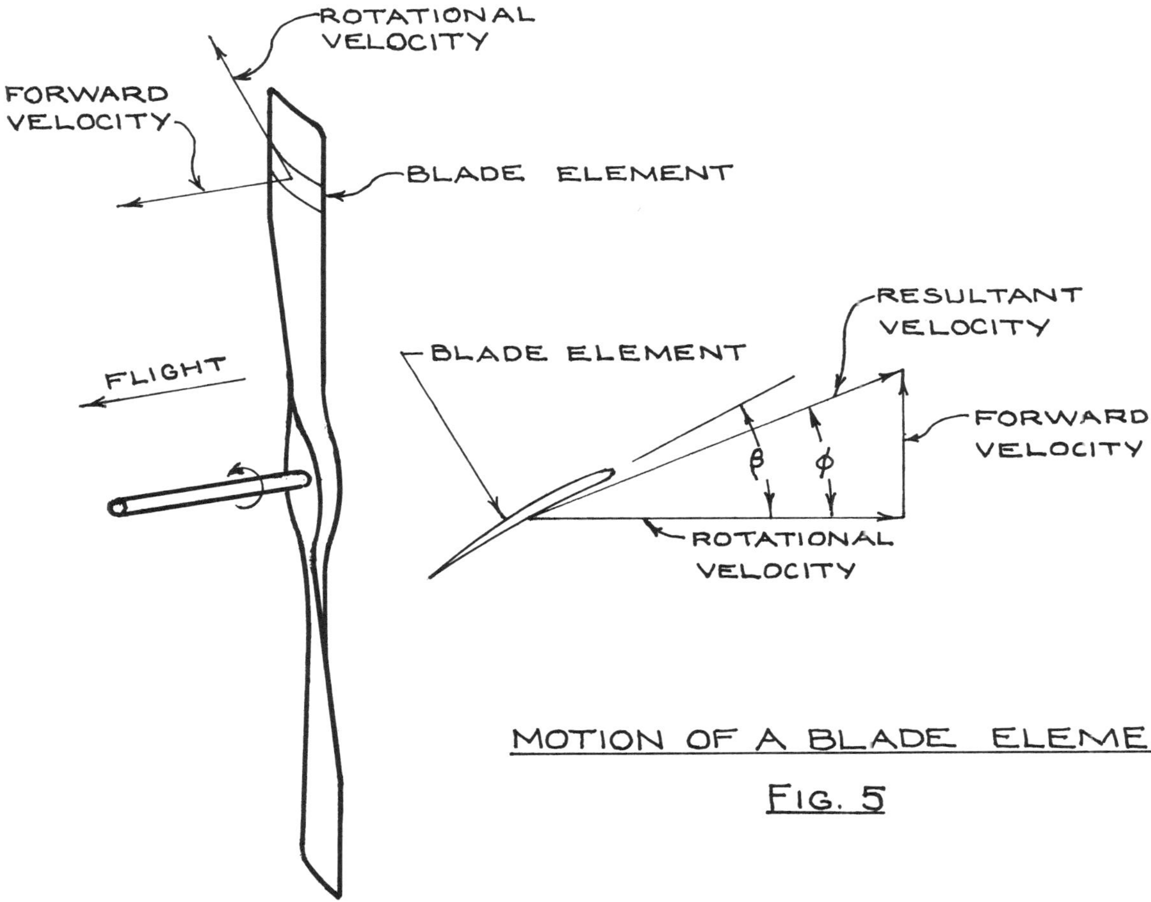

MOTION OF A BLADE ELEMENT

FIG. 5

rotate in opposite directions in order to cancel the torque which would have had a possibly disturbing effect on lateral trim.

The accompanying diagram, Fig. 6, illustrates the Wright propeller design system. The outline of an airfoil is a section of the propeller blade taken at a representative station along the length of the blade. The velocities, the angle of attack of the blade section, and the forces on it are calculated at this section as if the entire blade area were concentrated at this station. The computation was based on conditions at a station at 5/6 of the overall radius of the propeller. It is not known how they came to choose the 5/6 station as typical, but it is a pretty reasonable choice. More recent approximate studies of propeller performance have used 7/10 or 3/4 of the tip radius as a representative station.

The blade element in the diagram has a rotational velocity to the right and is advancing with the flight velocity, which is upward in the diagram. The apparent wind in the direction of flight is increased by the induced velocity (or slipstream), the flow produced by the action of the propeller on the air passing through it. It was computed by a simple formula based on momentum considerations. The relative wind at the blade element is the resultant of the three velocities, rotational, forward speed and induced velocity. It is shown at an angle ϕ_1 in the diagram. The blade element has a pitch angle β. The angle of attack of the blade element is $\alpha = \beta - \phi_1$. Knowing the angle of attack, the lift coefficient C_L and the drag coefficient C_D are found in the tables compiled in the wind tunnel tests of wings. The resultant force coefficient is $C_R = \sqrt{C_L^2 + C_D^2}$.

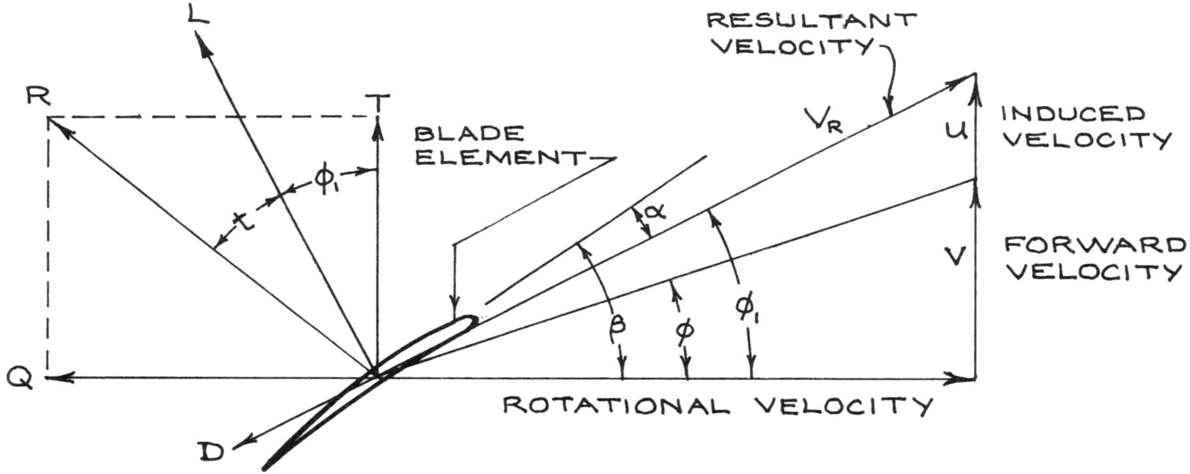

$C_R = \sqrt{C_L^2 + C_D^2}$

$t = \tan^{-1}(C_D / C_L)$

THEN $R = C_R (\rho S V_R^2)$

AND THRUST $T = R \cos(\phi_1 + t)$

TANGENTIAL FORCE $Q = R \sin(\phi_1 + t)$

WRIGHT PROPELLER THEORY

FIG. 6

The resultant force on the propeller blade is then

$$R = C_R \, k \, S \, V_R^2. \tag{4}$$

Here S is the total area of the propeller blades and V_R is the resultant velocity at the typical element. It is evident from the diagram that the thrust of the propeller is now just

$$T = R \cos(\phi_1 + t) \tag{5}$$

where t is the angle whose tangent is C_D/C_L. The tangential force is then

$$Q = R \sin(\phi_1 + t) \tag{6}$$

and the torque of the propeller is just the force Q multiplied by the radius to the typical element.

Today it seems crude to base the computations on a single blade element and it would be natural to compute conditions at a number of radial stations along the blade and then find the distribution of the thrust and torque forces from hub to tip. Having the distribution, the total thrust and torque would be obtained by numerical summation or a graphical integration. However, remember that the Wrights had no pocket calculator and probably no slide rule. All arithmetical computations were made by hand. This fact alone is sufficient to justify the simplified theoretical procedure which proved adequate for practical purposes.

The Wrights' formula for the induced (slipstream) velocity u may be derived as follows. Newton's second law states that force is equal to the rate of change of the momentum that results from application of the force. In the case of lumped masses, this results in the common formula Force = mass x acceleration. When applied to a continuous flow of fluid, the rate of change of momentum is not mass times acceleration, but rate of change of mass times velocity imparted. That is, instead of $F = M \times a$, we have $F = m \times \Delta v$ where m is the flow of mass per unit time and Δv is the imparted velocity.

The volume of air flowing through the propeller in unit time is A (V + u) where A is the disc area swept by the propeller, V is forward velocity and u is the induced velocity at the propeller. Air density is w/g where w is weight of air, lb/ft^3, and g is the acceleration due to gravity. The mass flow m is therefore (w/g) A (V + u). The velocity imparted to the air was assumed to be u, the induced velocity at the propeller. Consequently the momentum equation $F = m \times \Delta v$ becomes

$$T = (w/g)\, A\, (V + u)\, u \qquad (7)$$

Take the weight of air assumed by the Wrights, .075 $1b/ft^3$, and the acceleration of gravity g as 32 ft/sec^2. Then letting T/A be P, the average pressure rise through the propeller disc, we may write the equation in the form

$$u = \frac{426P}{u + V} \qquad (8)$$

With a slight change in notation this is a formula found in the notebooks of the Wrights and is apparently the formula used to obtain the induced velocity in their propeller design procedure. This is actually a quadratic equation in the unknown quantity u. That they wrote it in the form (8) in the early notebooks rather than as the quadratic solution of the equation (which appears much later) suggests that they had forgotten their high school algebra and probably solved (8) by a trial and error method.

In the early years they seem not to have realized that the velocity induced at the propeller is not identical with the final slipstream velocity far behind the propeller, but is half the ultimate slipstream velocity. They probably were aware of this from the works of Froude, but overlooked it in their calculations. The thrust is equal to the mass flow rate times the *final* velocity imparted to the slipstream which is twice the velocity u imparted at the propeller. Consequently equation (7) is incorrect and should read

$$T = (w/g)\, A\, (V + u)\, 2u \tag{9}$$

Then (8) becomes

$$u = \frac{213P}{u + V} \tag{10}$$

The difference between equations (8) and (10) represents a substantial error, but nonetheless great credit is due for the insightful attempt to combine the blade element and momentum theories. The fusion of these two theories, blade element theory and momentum theory, in a more developed form remained the most useful method of propeller design until the advent of high speed computers made modern vortex lattice computations practical.

Though account was taken of the backward flow of air induced by the propeller, there is also a rotational component, an induced swirl of the air, which the Wrights took no account of.

The analysis of the propeller problem outlined in the

foregoing pages is reconstructed from various notes and diagrams included in the published papers, (Ref. 7). It is a complete description of the aerodynamics of the propeller as understood by the Wrights, but there are no surviving notes indicating their procedure in applying the theory to the design of a propeller. It is not possible to start at any point in the given relations and proceed in a simple step by step manner to the design of a propeller. For example, if the desired thrust is assumed, one may calculate the induced velocity u, but then propeller rpm, blade angle β and blade area must be assumed and, by trial and error, a combination must be found which gives, finally, the value of the thrust originally assumed. Since thrust and torque can be computed, the efficiency may then be found, but undoubtedly a great number of combinations of the geometric parameters were tried before a practical and efficient propeller design was obtained.

In spite of some oversights and the rather drastic simplification of propeller aerodynamics inherent in the assumption that one element of the blade may adequately represent the whole propeller, the Wrights' propeller theory represents a notable advance in the understanding of propellers. They were the first, so far as I know, who combined the blade element theory with a reasonable attempt to define the relative velocity of the air at the blade using momentum principles. Their simplifications, though one might consider them crudities, are fully justified by their need for practical design information and the limited computational resources at their disposal. They had insight and were masters of economy of means to achieve their ends. And finally, they built propellers much more efficient than those of other experimenters of the time.

The Wright Engines

In their pursuit of the solution to the problem of flight, an engine was of secondary interest to the Wrights, but it was, of course, a necessity. They sent a letter of inquiry for a suitable engine to a number of manufacturers, but found none interested in meeting the stringent requirements for a small number of light weight engines. They very soon set about designing and building their own engine. In a remarkably short time, with the very able assistance of their machinist Charles E. Taylor, they produced an engine which met their needs for the first powered flyer.

In 1906 Orville Wright designed the four cylinder in-line upright water cooled engine which powered most of their machines through 1912. Weighing about 170 pounds, it produced about 35 hp at 1400 rpm. With this engine they found themselves in the aircraft engine business, a considerable number being sold to other aircraft builders.

The story of Wright engines is of subordinate interest to the history of their invention, but it is none the less remarkable. The interested reader is referred to reference (5) for details.

The Unique Risks of Inventing a Flying Machine

In committing themselves to the proposition that control was the primary problem that they must address, the Wrights were inescapably committed to flying their machines themselves at every stage as they developed them. This had also been Lilienthal's view of the problem of flight and he died in the course of his experiments.

Hiram Maxim, on the other hand, had confined his enormous steam powered flying machine to a track with constraints to prevent it lifting more than a few inches. During trials it broke loose and was immediately wrecked. Langley built large stable steam powered models. His hapless assistant was designated to steer his presumably stable man-carrying machine. Fortunately he suffered only a cold bath in the Potomac River when the machine suffered a structural collapse upon launching. Neither Maxim nor Langley had considered control to be the principal problem as had the Wrights and had exposed themselves to no personal dangers.

The Wright brothers realized that they had to learn the unknown art of flying, not in stable nor even adequately controllable machines, but would be simultaneously learning to fly and evolving a controllable machine. The threat of disaster was obvious. I can think of no earlier inventor taking a comparable risk. The steam engine, the steamboat, the cotton gin, etc. involved no necessary personal danger in their development.

The Wrights tried to minimize their risk by experimenting in the vast flat expanse of Kitty Hawk, flying against a substantial wind so that their relative speed over the ground was very low, and in early trials flew hardly more than a few feet above the sand. Most decisive, however, was their careful study of each strange behavior of their machine and an attempt at a remedy before proceeding with more flying. Their successful mastery of the art of flying as they developed the recalcitrant machine was a uniquely remarkable

7. The Fully Developed Wright Flyer.
Orville Wright Flying at Fort Meyer, Va. 1908

event in the history of invention.

Also extraordinary is the economy of means displayed by the Wrights at every turn. By one estimate their total expenditure on the development of the flying machine up to the first powered flight in 1903, including travel, was not more than $1,300. Langley, on the other hand, had quite generous grants from Congress and Maxim's project must have been similarly expensive.

There was always clarity and directness in their attack on each new problem. In contrast to the trial and error methods typical of the nineteenth century and of most of their contemporaries, the Wrights employed an analytic approach to the development of their "flyer", an early example of the methods of modern engineering.

Evolution of the Longitudinal Stability of Wright Airplanes

I have pointed out the rather serious instability of the early Wright machines in the longitudinal mode and that the Wrights became aware of this, but lived with the resulting demands on the skill of the pilot. In 1910 they seem to have reconsidered the basic configuration of their machines, possibly influenced by the success of various airplanes built by others in this country and in Europe. Through 1909 they were confident that they could not be seriously rivaled by others. In 1910 they were no doubt surprised that the world was rapidly catching up with them and, in fact, it was soon to leave them behind. The culmination of their re-engineering of the classic Wright configuration was the model B which appeared in 1911. It is interesting to consider the evolution of this type.

In order to understand what their experiments accomplished it is necessary to understand some simple aspects of longitudinal stability. First consider a simple aerodynamic device, the weather vane, in the form of an arrow with tail feathers. If it is subject to a steady wind, it will return to its original alignment with the wind after it is disturbed. It is a stable device; this is to say, when in a disturbed attitude it is subject to a moment, a tendency to rotate back to the equilibrium orientation where it points directly into the wind. If the pivot point is moved back nearer the tail feathers the restoring tendency will be less and it is clear that if it is pivoted too far back the force on the head of the arrow will cause it to turn away from the wind. It is then unstable. The pivot point where it becomes indifferent to wind direction before becoming unstable is the neutral point. For the weathervane to be stable in its intended orientation, its pivot must be ahead of the *neutral point.* An airplane, similarly, has a neutral point with regard to its rotation about a horizontal axis normal to its plane of symmetry, the pitch axis.

In considering the simplest aspect of stability, known as static stability, the center of gravity of the machine may be considered to be the point about which it rotates. Consequently for an airplane to be longitudinally stable, the center of gravity must be ahead of the neutral point.

In the early Wright machines there was a biplane elevator out in front and no horizontal surface behind the wings. In effect, the elevators are the arrow head and the wings are very large tail feathers. In the 1909 machines the resulting neutral point is estimated to be a bit less than a foot ahead of the leading edge of the wings. (See Fig. 8) The center of gravity (CG) of the machine is somewhere near one third of the wing chord back of the leading edge. (It cannot be far from this point since the principal weights, pilot, passenger, engine, fuel and wings are all close to this location.) Consequently the center of gravity is almost three feet behind the neutral point, an extremely unstable condition.

In 1910 several Wright machines were fitted with a single horizontal surface at the tail behind the rudder, the forward elevators remaining unchanged. The leading edge of the new tail surface was attached to the tail frame and was wire braced from the top and bottom of the tail frame to points near the mid chord of the surface. The trailing edge could be pulled up or down by control wires, curving the surface to produce a positive or negative lift. This surface was, in modern terminology, a combined stabilizer and elevator. The forward elevators and the aft elevator worked together in response to the elevator control lever. At least two machines of this configuration were present at the great air meet at the Indianapolis Motor Speedway in June 1910. In view of this it seems possible that at the time the Wrights thought of this as a new standard machine.

With the addition of a horizontal surface at a great distance behind the wings, the neutral point moved about a foot farther back. The machine remained unstable, but less so. This configuration turned out to be a short lived transitional type. One of these machines was flown experimentally with the upper forward elevator removed, and then with both forward surfaces removed.

In 1911 the model B appeared with no horizontal sur-

① CANARD WITH BIPLANE ELEVATORS (1909)
② STABILIZER/ELEVATOR ADDED AT TAIL (1910)
③ FORWARD SURFACES ELIMINATED, MODEL B (1911)

ONLY THE MODEL B HAS THE CG AHEAD OF THE NEUTRAL POINT AND IS LONGITUDINALLY STABLE.

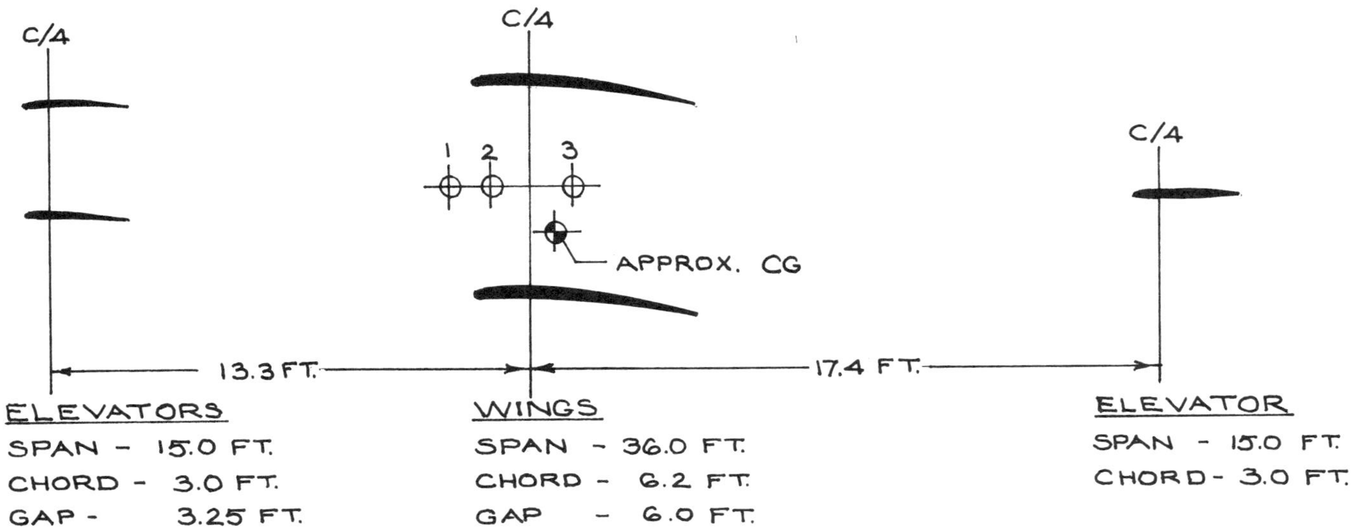

FIG.8 NEUTRAL POINT OF WRIGHT BIPLANES

9. The Transition Type, 1910.
Orville Wright demonstrating Turns for His Students at Huffman Prairie, June, 1910.

faces ahead of the wing, its longitudinal stability and control provided entirely by a single flexing surface at the tail. The elimination of the forward elevators brought the neutral point back more than two feet so that it was almost a foot behind the center of gravity. The new model B was statically stable and must have been much easier to fly than the earlier types. Unfortunately, so far as I know, there is no record of a pilot of Wright machines describing the difference in handling characteristics of the original Wright machines and the model B. Charles Wald, father of the author and one of the pilots trained by the Wrights, considered the model B a pleasure to fly, but he never flew a machine of the earlier type.

While the Wrights certainly must have realized what a great improvement of stability and ease of handling they had brought about as they evolved the model B, they did not think of stability as we now do. They never seem to have thought about the change of pitching moment as the angle of attack changes, the essential aspect of longitudinal stability. Also neutral point is a relatively modern idea which was not known to them.

10. The Wright Model B, 1911

Part 2
Flying with the Wright Brothers, One Man's Experience

The Young Enthusiast

In 1905, Charles Wald, a young man in the employ of the U.S. Customs Service in the Philippines, came upon an article describing successful flights by the Wright brothers and illustrated with photos of Wilbur and Orville Wright gliding at Kitty Hawk. He was immediately excited by the possibilities of mechanical flight, thinking "That's for me!" and determining to become a part of the new human adventure. In 1906 he returned to the United States and, on his way east, stopped in Dayton to visit the Wrights and learn what he could of their experiments. To his surprise, they invited him to their home where he met their father, Bishop Wright, and sister Katharine. He must have tried to impress the Wrights with his desire to be associated with them, but apparently received little encouragement except advice to learn something of engineering. On returning to New York and continuing employment with the Customs Service, he enrolled in night courses in engineering at Pratt Institute.

While employed at the New York Custom House during the next several years his enthusiasm for flying was such that he attended every flying event that came to his notice. He was present at Governor's Island in September 1909 when Wilbur Wright made his spectacular flight up the Hudson to Grant's tomb and return, the flight in view of many thousands of New Yorkers that brought America to the realization that mechanical flight had become a reality. During

Wilbur Wright's preparation for that flight, Charles Wald saw him land after a short time in the air, saw a bit off the tip of each propeller blade, rasp it fair and smooth, and then take off again. This quiet confidence and resourcefulness impressed Wald and all those who knew Wilbur Wright.

Wald attended the first air meet and public flying by the Wright exhibition team at the Indianapolis Motor Speedway in June 1910. He was at Simms Station near Dayton the same month, at a meet at Asbury Park, New Jersey in August where the Wright model B made its first appearance, at another meet in Mineola, New York, also in August, and at Belmont Park, Long Island in October. In 1909 Charles Wald and a number of friends formed the Aero Club of Long Island. One of the projects of the club was the construction and flying of a Chanute type glider.

Learning to Fly

Wald apparently was first rewarded for persistent requests for employment with the Wright Company in 1911 when he was employed in the factory in Dayton. Apparently there was insufficient need for his services later in the year and he was let go. This is attested by a letter in which he reminds Orville Wright of his offer to re-employ him "when business warranted". In January 1912 he sent several telegrams and letters from New York to Orville Wright negotiating for re-employment. In a letter dated March 18, Orville Wright offered employment in the shop, stating that wages were 25 to 35 cents per hour. Perhaps there was a change of circumstances or Orville Wright may

11. Charles Wald at the Controls of a Model B.

This machine, photographed at Huffman Prairie in April, 1912, is evidently one that was converted from the earlier configuration with forward elevators and is equipped with dual controls for student instruction. The loose wire above the pilot's head was a control which released spring loaded shims under the valve stems, permitting instant engine stopping and compression release, thus allowing the propellers to spin freely during the glide approach to a landing.

have been impressed by Wald's enthusiasm. In any event, Wald began flight training as an employee of the Wright Company on April 12 at Huffman Prairie, a pasture eight miles east of Dayton at Simms Station, a stop on the suburban electric car line from Dayton.

Charles Wald was my father and from him I have inherited many photographs of those early flying experiences, his log book, and a few old news clippings. I also have memories of some of his accounts of his flying in the Wright machines and of the men who flew with him. Since the "hydroaeroplane" operation at Glenwood Landing, in particular, seems largely forgotten, I am trying to fill in the record as completely as I am able.

His flight training began with instruction by A. L. Welsh in a machine which had been of the original Wright type and later converted to the Model B configuration. All Wright machines before 1910 had biplane elevators ahead of the wing, what we would now call a canard configuration. The Model B, introduced in 1911, had a single horizontal control surface attached to the tail frame behind the double rudder. The conversion is evidenced by the structural member, identifiable in the photographs, which extended forward and upward from the leading edge of the lower wing. This vestigial member was originally part of the structure supporting the front elevators and is absent in machines built originally in the B configuration. By this vestigial feature, a number of machines identified in photographs as Model B may be recognized as conversions of older types. In all the Wright machines after 1905 the pilot and passenger (or student) sat side by side on the lower wing over the front wing spar.

Wald was instructed by Welsh every day that weather permitted excepting Sundays, which were accorded religious respect by the Wrights, until his primary training was considered complete on April 23. The Wright machines had what would now be considered an extremely light wing loading of about two and a half pounds per square foot, which made them very sensitive to gusts and turbulent air. Flight instruction began very early in the morning when the air was still and usually ended when the sun's rays had begun to warm the field and thermal air currents became established. Flying was usually resumed after five PM when the air was again still.

The students spent many hours in the shop working on the engines and making repairs during the time between morning and evening flying. In the training at Huffman Prairie, the Wright pilots acquired extensive experience in repairs of every description. Since student pilots were only permitted to fly when there was practically no wind, during most of the hours of the day they were kept busy in the shop learning to maintain and repair their machines. Breakage of propellers, wing ribs, and skids was frequent. The rollers on the propeller drive chains often broke and various engine failures were common. By the evidence of Wald's log book, after what seems to us rather serious damage the machine is often flying again in a few days.

The log entry for Friday, April 12th reads:

> First lesson 9 minutes. A. L. Welsh instructor remarks "good work" after first flight handle both levers, second flight 3 minutes duration, 2 errors in warp on turning, Welsh advises practice on balancer. Conditions ideal no wind.

Orville Wright had devised what we would now call a flight simulator, the world's first, and this is the "balancer" referred to in the log. It was an old airplane balanced on a trestle in the factory. The lateral control was connected to clutches at the wing tips which could engage continuously moving belts. This primitive mechanical simulator provided the student pilots a means of practicing the use of the very unnatural lateral control system used by the Wrights. Moving the control lever forward brought the left wing down, moving it backward depressed the right wing. Charles Wald's log book indicates two hours spent on "the balancing machine" on Sunday, April 14, and again two hours several days later when the wind was too strong for student flying. The balancer is also mentioned on other days. No photograph of this contraption has been found.

On Tuesday, April 16th:

> Get up 5 A.M. brisk wind no flying. Bergdoll finishes with 10 minute flight in P.M. Capt. Baldwin and Orville Wright on field, Baldwin refuses flight with Orville on account of wind. Orville has fun at Capt's expense.

The log entry for Saturday, April 20th reads:

> Get up 5:00 AM. 6 o'clock car to Sim's. Flight 8:10 AM 14 minutes duration puffy light wind. Figure eights. PM three flights ... 9, 20 and 17 minutes respectively, right and left turns and 8s. Ran into wake of machine on short turns some lively warping to overcome roll and pitch of machine. Start, get-away and landings. 1st landing OK. Total time for day one hour. flying OK.

On Tuesday April 23rd the log concludes:

> 6 landings rather hard, slight improvement at the finish. Last landing cut off power at 100 ft. altitude. Welsh thinks no more time necessary. Start for New York 8:50 PM.

Primary flight instruction was apparently considered complete after a total of fourteen flights on five days for a total of two hours and forty-six minutes in the air. The new pilot returned to New York the same evening without having actually made a solo flight.

On May 30 Wilbur Wright died of typhoid fever and subsequent events confirm that the spirit of the Wright team and the Wright Company died with him. Charles Wald felt the death as a great personal loss and the passing of a genius. On June 11th his former instructor, A. L. Welsh, and Lt. L. Hazelhurst of the Army Signal Corps were killed in the crash of a Wright machine at College Park, Maryland.

A letter dated June 12, 1912 reads in part,

> My dear Mr. Wright:
>
> Lately you have had more sorrow and anxiety than is usually allotted to the average man in a lifetime and I have felt the loss and pangs of grief as much as anybody possibly could outside of your immediate family, consequently I feel a timidity in broaching a subject on material things and business. Realizing that my late friend Mr. Welsh's place must be filled and the work must go on, I take the liberty of writing you on this rather deli-

cate subject reminding you that in the past you have promised me consideration in case of a vacancy....

Very respectfully yours,

Chas Wald

A terse letter to Charles Wald at the New York Custom House, dated June 19, 1912, concludes:

> In our talk over a man to take Mr. Welsh's place, you were given special consideration, so that I think it would be well for you to see Mr. Barnes about the matter.
>
> Very truly yours,
>
> Orville Wright

On June 27th Charles Wald was at Huffman Prairie as an employee of the Wright Company and completed his first solo flight. The machine in which he flew that day was the old B-14 which had been modified from the front elevator configuration, but now rebuilt to the true "B" configuration. It had not been flight tested since the overhaul and the new pilot quickly discovered that it was badly rigged so that it tended to roll to the right and the control wires were too tight.

The Wright four cylinder engine had no throttle control, only a spark retarding device operated by a foot pedal which provided a measure of power reduction. On attempting to reduce power, Wald found the control inoperative. Since the engine was on the wing at the pilot's right hand, he was able to try to free the spark retard. While his attention was thus engaged he suddenly realized that he was rapidly approaching two large trees with a rather narrow space between them.

12. Wald and Welsh in Flight Over Huffman Prairie, April, 1912.
This photo provides a clear impression of Huffman Prairie with the fine old trees which must have been something of a hazard for student pilots.

With what seemed to him a violent maneuver, he passed between the trees. Not surprisingly, the log concludes "...rough landing. Time 5 PM." So ended the ten minute solo flight.

On the following day he flew for fifteen minutes with Orville Wright in the seat beside him checking his competence. "Mr. Wright makes adjustments on machine."

From that time until August 6th he flew every day except Sundays and two days spent in the shop when high winds prevented flying. From the solo flight on June 27th he had made seventy four flights and had spent thirty-two hours and thirty-eight minutes in the air. On most of these flights he was accompanied by another student, William Kabitzki, but it is not clear from Wald's log how time at the controls was divided between them, if at all..

There was some horseplay among the student pilots. One of them, Harry Atwood, apparently as a result of a somewhat arrogant attitude toward the others, became the butt of practical jokes. An instance is depicted in a photograph in which he is whitewashing a line across the field with a broom while others are grinning behind his back. He had been told that he would need the line as a guide to learn to fly straight. Another leg pulling required Harry Atwood to develop his sense of balance by standing on the axle of a two-wheeled cart, balancing himself as long as possible while one of the others kept time with a stop watch. "Mr. Wright finally put a stop to this."

On days when the weather prevented flying and the need for mechanical repairs was not pressing, there were often poker games in "the shed", the hangar constructed at

13. Harry Atwood Whitewashing a Line Across the Field.

He was the victim of a number of practical jokes. Here he is whitewashing a line which he had been told was a necessary guide to learn to fly straight. The amused onlookers are, left to right, Howard Gill, Clifford Turpin, T. D. Milling, H. H. Arnold and A. L. Welsh, Huffman Prairie, June, 1911. "Hap" Arnold became the commanding general of the U. S. Air Force in World War II.

14. Atwood Developing His Sense of Balance, June 1911.
Watching the performance are H. H. Arnold, A. L. Welsh, T. D. Milling, Orville Wright, Oscar Brindley and Calbraith Rodgers who is acting as Timekeeper.

15. A Poker Game in the Aeroplane shed, Huffman Prairie, 1911.
Left to right, Clifford Turpin, T. D. Milling, Oscar Brindley, John Rodgers.

16. Charles Wald at the controls of a new Model B, Huffman Prairie, 1911.

The Wrights usually used only one flight instrument, a piece of string tied to the cross bar between the front ends of the landing skids. It can be seen in this photo hanging below the front end of the near skid. By streaming to one side or the other, this simple and versatile device informed the pilot that he was slipping left or right. If he flew too slowly and close to a stall, it described a high arc. In normal flight it streamed straight back.

Huffman Prairie in 1910.

One may reconstruct in imagination those hot summer days, the complex clattering machine with pilot in cap and shirt sleeves skimming in wide turns low over the Ohio prairie around a clump of ancient trees. It is more difficult to imagine how strange the scene was to occasional visitors to Huffman Prairie who had never seen a flying machine. Perhaps less difficult is the appreciation of the state of mind of those eager young men, largely unaware of the hazards they faced, some with dreams of fame and glory, others with a more sober hope of becoming a part of a great new human enterprise.

Excerpts from the log book suggest how unreliable the machines were and how dangerous flying was at the time. The frequent breakage of chain rollers and engine parts is a consequence of the inferiority of the materials then in use compared with those available a few decades later. The following are excerpts from Charles Wald's log of flights at Huffman prairie during that summer of 1912.

June 29	Replaced broken roller in chain.
July 2	Stay wire of horizontal rudder broken at eyelet.
July 3	Lost pin holding exhaust rocker arm causing missing cylinder,
July 4	Skid broken by rough landing in hummocks, break showed skid partly rotted. New skid fitted in two hours.
July 6	Rear truss wire at skid broke at loop in leaving ground breaking rib and nicking propeller.

July 8 Propeller broken by running machine into shed under power, machine running off run way and striking floor. Replaced with spare propeller.

July 16 ...replaced broken roller in chain.

July 20 Connecting rod #2 cylinder broke at bronze casting below wrist-pin bearing, breaking entire piston and cam shoes, the obstruction jamming in crank case, tearing hole in crank case .... Altitude at time about 300 ft. (No activity on 21st, which was a Sunday.)

July 22 Take #19 motor to field & install in machine B-14. Change propellers. (Flying was resumed at 5:15 P.M.)

July 27 Two rollers broken in chain, replace spark plug ... clean and adjust magneto.

July 29 Crank case cracked at cam shoe bracket. Motor operating OK. Take motor down and patch crank case.

July 31 12:30 PM run into rough ground with wind astern strike hummock with wing tip. Three ribs and forward spar broken. PM replace broken wing.

Aug 3 13 rollers broken in short chain, 2 in long chain. 3 ribs broken by taking machine into shed under power. Kabitzki operating.

Since these machines had no means for steering on the ground and, of course, no brakes, it is remarkable that twice the log mentions bringing the machine into the shed by its own power. It is not surprising that both instances are followed by mentions of breakage!

On August 6th Orville Wright again flew with Wald on a ten minute flight.

In ordinary conservative maneuvers, the model B biplanes were, from an aerodynamic point of view, reasonably safe, but attempts at spectacular maneuvers and especially altitude records were extremely dangerous for a reason not then appreciated. While the Wrights were fully aware, in consequence of their wind tunnel experiments on lifting surfaces, that there was danger of wing stalling, i.e. loss of lift at high angles of attack, it was some years before the resulting uncontrollable and fatal spin was recognized. Two of the Wright pilots, Ralph Johnstone and Arch Hoxsey, died in 1910 after reaching what were then considered high altitudes where a stall was likely and then entering a "spiral dive" from which they did not recover.

By mid 1912 five Wright pilots and two U. S. Army men had been killed in Wright machines. Aside from spins, there were structural failures and probably one or more disasters which might have been avoided by the simple and, one would think, obvious expedient of safety belts.

The Hydroaeroplane School

It is frequently stated that Wilbur and Orville Wright first experimented with water based aircraft in 1913. They conducted experiments with planing hulls driven by air propellers on the Miami River near Dayton in 1907, but at that time no aircraft was flown off the water. McFarland (Ref. 7), states

The CH was the first Wright hydroplane. Essentially it was a Model C to which pontoons had been added for the purpose of taking off and alighting on lakes, rivers, and inland waterways having calm or shallow waters. In experiments made on a secluded stretch of the Miami River near Dayton in the spring of 1913, two "multi-step" pontoons, 16½ ft by 1½ ft, were attached to the skid runners. It was found, however, that the long twin pontoons interfered with the control of the machine in the air, particularly during short turns. The fully developed CH therefore used a single wooden 240-lb pontoon, 10 ft by 6 ft by 10 in. A small pontoon supported the tail.

It is also stated by Crouch (Ref. 3) that his (Orville Wright's) first successful water takeoff and landing were made with a Model CH on a secluded stretch of the Miami River in 1913.

However, there is evidence that three pilots trained by the Wrights, Frank Coffyn, William Kabitzki, and Charles Wald, were doing a lot of flying with model B flyers mounted on floats in 1912 and of course Orville Wright must have flown this type for its initial tests.

Charles Wald, in a letter to Orville Wright dated March 12, 1912 describes the flying of Frank Coffyn. This is the first reference to the flight of a Wright machine on floats that I have come upon. The floats on Coffyn's machine were not of Wright manufacture.

"Coffyn is doing good work over New York Bay almost daily. The "get-away" of the machine

> is somewhat lengthy, the distance I judge to be about one hundred yards. To use a nautical term, the floats carry a "heavy bone" or in other words do not shear the water very cleanly. Except for a shear at the bows the floats are square. In the air the machine acts very fine except for a slight pendulum action but this may be due to Coffyn's own peculiar way of flying".

At that time Wald was working at the custom house near the foot of Manhattan Island. There is a photograph, of Coffyn's "hydroaeroplane" afloat in the boat basin near the custom house with the liner Olympic, sister ship of the Titanic, in the background steaming down the Hudson

In the photographic history "Kitty Hawk and Beyond" (Ref. 4) there is a photo entitled "Rescue of William Kabitzki After Crash at St. Louis, Missouri, 1912". All that is visible of the wreck floating in the river is the bottom of a pair of floats of the same multi-step design that MacFarland describes.

It seems to be a forgotten chapter in the history of the Wright Company that they established a school of water flying, occasionally referred to as the Wright air ferry, at the Glenwood Country Club at Glen Head, Long Island, N.Y. in 1912. This operation was established on the shore of Long Island Sound in a region of great yachting activity where there were many estates of the very wealthy. It was hoped to bring to the attention of yachtsman and other wealthy people the possibility of water flying as a sport and to offer flying lessons to that end. Charles Wald was put in charge of this operation.

The machine used was a Wright Model B, designated B-9, powered with the four cylinder water cooled Wright engine which was used in most Wright machines from 1908 through 1912. This engine developed about 35 horsepower at 1500 rpm. The machine had a gross weight, with floats, of about 1300 pounds and had a wing area of about 460 square feet. It was considered an old machine, having been converted from the older Wright configuration to the B configuration. B-9 was fitted with a pair of wooden floats with three steps, probably identical with the floats described in the above quote from McFarland (Ref.7). It was also provided with a water rudder which was attached to the air rudder and hung below it. The rudder is clearly visible in several of the photographs.

Charles Wald had never flown the hydroaeroplane when he arrived at Glen Head. Almost certainly Orville Wright would have test flown the machine before it was sent from the factory in Dayton to Long Island. Wald made the first flight at Glen Head early on the morning of September 9, 1912 in a harbor where four masted schooners still plied their trade carrying lumber, sand and gravel under sail. From the log: "Machine leaves water in 100 to 200 yards, little spray, smooth sea, ...". The flight lasted thirty-five minutes. In another flight at dusk on the same day with the water very smooth, Wald was unable to judge his height above the surface and struck the water in a turn. Two elevator spars and four wing ribs were broken. The difficulty of judging height above smooth water became well known to marine pilots, but Wald, with no instruction in water flying, was discovering the special art of water flying for himself. After the damage incurred in striking the water, Wald was flying again eight days later.

17. The Hydroaeroplane in the Boat House at Glenwood Landing, Sept. 1912.

18. The Hydroaeroplane on the Launching Ways.

19. The Hydroaeroplane Being Launched.

20. The Hydroaeroplane Taking Off.
Here the age of flight and the age of sail overlap. Four masted schooners were still carrying cargo on Long Island Sound in 1912 and for some years afterward.

On September 21, having acquired more experience with the machine, Wald flew from Glenwood landing to New Rochelle Harbor, nine miles across Long Island Sound, carrying a token cargo of newspapers in a twenty-five mile per hour wind. This flight evoked considerable public interest and was greeted enthusiastically in New Rochelle Harbor by horns and whistles of harbor craft. He made the return flight the following day, a Sunday, and earned a reprimand from Orville Wright for flying on the Sabbath.

The next day one J. C. Jackson became the first passenger of the "hydro". The log records "...machine leaves water with an additional run of about 50%. Plenty of reserve power in air." With the extra weight of a passenger and the weight of the floats, it is remarkable that this machine with a 35 horsepower engine was able to get off the water at all. That it could do so is a tribute to, among other things, the very efficient Wright propellers.

Following this flight some changes were made in the position of the floats. It is evident that the floats provided marginal buoyancy and there must have been considerable risk of plowing in and even pitching end over end in choppy water. It was probably at this time that Orville Wright attached a small canvas lifting surface near the bow of each of the floats, presumably to plane on the water surface and obviate the danger of plowing in. However, when Wald took off, he very nearly came to disaster due to the severe nose up trim caused by the aerodynamic lift of these surfaces. They were removed before the next flight.

On several occasions Wald was impressed with the great strength of the rising air currents he encountered when, after flying over the water, he found himself over the sun baked sand pits that abound in the area. A strong thermal must have caused a considerable jolt since the wing loading of the "hydroaeroplane" was only about 3 pounds per square foot. He may have made some attempts to soar with power off. All the Wright machines were equipped with a compression release which unseated the exhaust valves, cutting the power and permitting the propellers to rotate freely with minimum air resistance in gliding flight.

On October 4th a second passenger, C. G. Goddard, was carried in a twelve minute flight. On the following day Wald flew to Larchmont, then to New Rochelle, and returned to Glen Head with a flying time of thirty-five minutes in calm air.

On October 10th while preparing the machine for flight, Wald became aware of a man in distress, he having fallen overboard from a rowboat some distance offshore. The machine was launched and, barely skimming the wave crests, quickly reached the exhausted man who was pulled aboard and put in the passenger's seat. This event was quickly picked up by the newspapers, who undoubtedly were keeping an eye on the flying activities, and was widely reported. A photo of a re-enactment of the rescue was printed and several rather lurid Sunday supplement articles, trumpeting the advent of aerial rescue, followed.

The evening of the same day Marion G. Peck, a newspaper reporter, was carried as a passenger. As fog gathered during the evening chill, the magneto became saturated, causing the engine to miss. Flying low before a light breeze with the extra load and a missing engine, the "hydro" was unable to rise high enough above the water to turn into the wind and the pilot attempted a downwind landing. The floats plowed in and the machine flipped over on its back. Wald was briefly trapped in the water amid the wire bracing. When he surfaced, the indomitable Miss Peck was already seated on the bottom of a float, saying "I was afraid you were never coming up!". The wet and chilled pair sat on the floats of the capsized machine for several hours in the gathering darkness before being picked up by a passing fishing boat.

This eventful day concluded the activities of the Wright air ferry for the season and for all time. The "hydro" had made thirteen flights and spent a bit over five hours in the air, although it is possible that the log on which this statement is based is incomplete. Wald claimed to have reached an altitude of 4500 feet in this machine, which was, in 1912, an unofficial record for a water based craft. Though several passengers were carried and a token cargo of newspapers was carried on several occasions, there is no record of any flight instruction being given. The failure to attract students may have been the reason why Orville Wright abandoned the operation at Glen Head.

It seems doubtful that old B-9 would have been made flyable again after the capsize and the abandonment of the flying school at Glen Head. However, in the Wright archives at Wright State University in Dayton, Ohio, there is an excellent photo of a machine in the Wright factory which can be positively identified as the machine flown by Wald at Glen Head. The photo is dated 1913 and otherwise unidentified. It seems probable that it is misdated and actually shows the machine in preparation for transporting to Long Island in 1912.

When the hydroaeroplane school at Glen Head was abandoned at the close of the 1912 flying season Wald was let go by the Wright Company. On leaving the Wright Company, he, with several friends, formed the short lived Manhattan Aeroplane Company which, in 1913, built two or three airplanes, one of which was a flying boat. During the first World War he was head of the inspection department at the Curtiss factory in Buffalo. In 1923 he was in charge of the victorious Schneider Trophy seaplane racing team sent to England by the U.S. Navy. He continued his work in aviation with the naval service until his retirement in 1947.

21. The Hydroaeroplane in the Wright Factory, Dayton, Ohio.

References

1) Ira H. Abbott & Albert E. von Doenhoff, Theory of Wing Sections, Dover Publications, Inc., New York, 1959.

2) Octave Chanute, Progress in Flying Machines, Dover Publications, Inc., New York, 1997, originally published 1894.

3) Tom D. Crouch, The Bishop's Boys, A Life of Wilbur and Orville Wright, W. W. Norton & Co., New York, 1990.

4) Ronald R. Geibert and Patrick B. Nolan, Kitty Hawk and Beyond, The Wright Brothers and the Early Years of Aviation, A Photographic History, Wright State University Press, Dayton Ohio, 1990.

5) Leonard S. Hobbs, The Wright Brothers' Engines and Their Design, Smithsonian Institution Press, Washington D.C., 1971.

6) Sighard F. Hoerner, Fluid-Dynamic Drag, published by the author, Midland Park, New Jersey, 1958.

7) Marvin W. McFarland, editor, The Papers of Wilbur and Orville Wright, McGraw-Hill Book Co., Inc., N.Y., 1953.

8) Charles Wald, Flying Report, The Wright Company, 1912. Collection of the author. Copy in the collection of the Dunbar Library, Wright State University, Dayton, Ohio.

9) Howard S. Wolko, ed., The Wright Flyer, an Engineering Perspective, Smithsonian Institution Press, Washington D.C., 1987.